Luciano J. Alvarenga

The Cerrado biome under the law

Luciano J. Alvarenga

The Cerrado biome under the law

ScienciaScripts

Imprint

Any brand names and product names mentioned in this book are subject to trademark, brand or patent protection and are trademarks or registered trademarks of their respective holders. The use of brand names, product names, common names, trade names, product descriptions etc. even without a particular marking in this work is in no way to be construed to mean that such names may be regarded as unrestricted in respect of trademark and brand protection legislation and could thus be used by anyone.

Cover image: www.ingimage.com

This book is a translation from the original published under ISBN 978-3-330-75136-1.

Publisher:
Sciencia Scripts
is a trademark of
Dodo Books Indian Ocean Ltd. and OmniScriptum S.R.L publishing group

120 High Road, East Finchley, London, N2 9ED, United Kingdom
Str. Armeneasca 28/1, office 1, Chisinau MD-2012, Republic of Moldova, Europe
Printed at: see last page
ISBN: 978-620-8-09214-6

Chapter 1

INTERDISCIPLINARY PROBLEMATISATION

1.1 *First movement of ideas*: Environment and History

Brazil is the land of "megadiversity" (Brandon *et al.*, 2005). With five biomes and a gigantic river network, Brazil's continental expanse harbours the richest and most diverse biota on earth. The Amazon basin is characterised by the highest degree of terrestrial and freshwater biodiversity at both national and global levels. In addition to the Amazon, Brazil has two *hotspots*, so named because they are considered priority regions for biodiversity conservation: the Atlantic Forest and the Cerrado. The country also has the largest tropical wetland in the world, the Pantanal.

It should also not be forgotten that the country is also known for its geological diversity, which can be seen in countless geologically interesting places in all parts of the country (Silva, 2008).

[2]However, especially since the colonial period, Brazil's natural systems have been intensively exploited (Pádua, 2004a). A typical literary account of the pattern of territorial occupation that began at this time can be found in the treatise on the Brazilian economy published in 1711 by the Jesuit André João Antonil, in which he wrote: "Once the best land for the sugar cane is selected, it is ploughed, burned and fed, removing everything that could serve as an obstacle".

The logic of this practice, based on the view of native woods and forests

---

[1] Fernandez (2005, p. 5) cites a number of examples that call into question the so-called "ecological crisis" as a recent phenomenon. Referring to the history of Brazil, he recalls that when the Portuguese officially landed in 1500, "the 'virgin' Atlantic forest was already almost entirely secondary and (at least in the flat areas) completely destroyed by the coivara agriculture practised by the Indians". As Fernandez explains, "swidden farming" is an agricultural method developed by the Indians and used by the colonisers. This technique involves creating clearings in the forest by removing the vegetation. The piles of dried shrubs - "coivaras" - are then burnt so that the ashes can be used as fertiliser. However, the arrival of the European colonisers in Brazil and, furthermore, the progressive spread *of northern* world views in the tropical colony meant a progressive geographical expansion and intensification of environmental degradation processes.

as "obstacles" to the free development of the forces of extraction and production, prevailed in the following centuries and was accompanied by social injustices. In his 1883 work "O abolicionismo" (Abolitionism), Joaquim Nabuco denounced the "breath of destruction" resulting from the perverse combination of environmental destruction and slavery, which led him to lament: "At every turn we find and feel the traces of this system that reduces a beautiful tropical land to the appearance of regions where the creative power of the earth is exhausted...".

Many years have passed since the writings of Antonil and Nabuco. However, the carelessness that characterised the occupation of *Terra Brasilis*, which was seen as an infinite supply of natural resources, is still present in the mentality and behaviour (Pádua, 2008). The motivation to conquer, appropriate and exploit new territories is at the root of the land, and this motivation has led to an extensive, spatially intensive pattern of land use in relation to natural resources (Moraes, 2005). "In few countries", says Pádua (2004a), "is the weight of the past as great as in Brazil".

However, it would be foolish to reduce the causes of the country's socio-ecological crisis to its colonial past. The negative effects of the extensive monocultures and slavery, the extraction of mineral resources and other exploitative activities that took place here were not recognised by the majority of the population at the time, although there were some who denounced them, as Pádua (2004b) noted. An exploitative colony was and is always "a brutal and immediate endeavour", and it would have been naïve to expect, in the case of Brazil, that the colonisers would have been guided by a long-term development model typical of "the idea of the nation, the ideal of the historical continuity of a political community" (Pádua, 2008).

1.2 *Second movement of ideas*: the utopian proposal

What needs to be questioned today is not Brazil's colonial past, but the fact that the plundering of natural resources has been the predominant *form of* land use throughout the history of the independent country. As Pádua (2008) states, building a true nation requires "a new logic based on the care and preservation of the ecological, social and cultural foundations of collective existence". For many scholars, the "new logic" to which Pádua alludes is synonymous with the proposal of "sustainable development", whose origins date back to the 1987 report "Our Common Future", the result of the work of an international commission led by Gro Harlem Brundtland. Since then, the term has been used to summarise styles of development that aim to meet the needs of the present generation without compromising the ability of future generations to meet their needs. This means enabling people today and in the future "to achieve a satisfactory level of social and economic development and of human and cultural fulfilment, while making appropriate use of the earth's resources and conserving species and natural habitats" (United Nations, 1987).At a theoretical level, however, the proposal has been subjected to semantic distortions for convenience, depending on the hidden interests of various interlocutors, actors and market companies. They label themselves as "sustainable", sometimes with wide media coverage (*greenwashing*), and in reality are far from it. This confirms the conclusion that the *idea of* sustainability, originally intended to bring about economic and socio-ecological change, is often hijacked by companies - not coincidentally those that destroy the environment the most - in order to camouflage and legitimise the continuation of practices that are harmful to nature.

What is certain, however, is that the concept of sustainability continues to linger in an abstract space of different ideologies and concepts and contributes little to the concrete reinvention of production and consumption processes.

For this reason, the concept is heavily criticised, for example by Fernandez (2011, pp. 193-194), for whom the proposal of sustainable development is an *oxymoron* because it

> [...] it does not escape the laws of the market and the traditional economy [...] *the system only works if it grows.* Any growth achieved through a "sustainable" development project, by growing the economy, *creates the need for more and more growth.* We then fall into the same trap. Sustainable development is an *oxymoron*: in a finite system, every

process is either development or sustainable. 'Sustainable' development, if it creates a greater awareness of the need not to overexploit resources in the short term, can at best buy a little time (that is, of course, if it is not first destroyed or replaced by endeavours that set no limits to its own growth). But that would only postpone the problems; much more profound changes in our economic philosophy are needed to solve them. On the other hand, "sustainable development" is also an extremely dangerous concept, because it gives people the comfortable illusion that sufficient measures are being taken to preserve nature and that this would be possible without fundamentally changing the logic of the economy. For some, both producers and consumers, "sustainable development" is a term that serves wonderfully to soothe their own conscience. For others, who are more cynical, it is a fabulous commercial strategy to get their products into a market that is increasingly concerned with environmental issues. Whether it fulfils any other purpose beyond that is, to put it mildly, highly debatable.

Ultimately, it appears that many use the argument as an evocation of a utopia (Veiga, 2008), as formulated in the report "Our Common Future" (United Nations, 1987); however, few are in a position to know and experience the sacrifices and profound changes that the concrete adoption of truly sustainable lifestyles would entail.

The fact is that despite its good intentions, in practice the proposal has not been able to persuade economic, governmental and social actors to change their behaviour patterns. [2]The willingness for consensus and sacrifice is still low when it comes to actually experiencing sustainability in the individual and collective sphere.

1.3 *Third movement of ideas*: between old and new theories of knowledge

But we need to move forward, and along the way, there are those who reserve to science the task of helping to ensure that natural systems can be utilised by humans in a truly sustainable way. The signatories of this thesis assume that in an ideal world, scientific knowledge will be at the centre of guiding human projects and actions.

However, this statement must not be taken to extremes. To accept it unreservedly would be to endorse a subtle form of authoritarianism resulting from the predominance of "Eurocentric" rationalism (Dussel, 1993), which is interventionist, mechanistic and fragmentary, to the detriment of other, sometimes more coherent and comprehensive, ways of understanding reality (Feyerabend, 1991; Santos, 2001; Morin, 2001). However, from the point of view of the protection of natural systems (ecosystems, watersheds, biomes, etc.), the importance of scientific knowledge seems indisputable. In a *dialogue of knowledge*, in the open and participatory space of interdisciplinary epistemologies, the sciences dedicated to nature conservation can contribute to the improvement of legal and political environmental standards (Leff, 2002).[3]

For this to be possible, the work dedicated to the realisation of the idea of sustainability must move from the abstract and symbolic space of discourse - often ambiguous and characterised by the current ideology of economic growth at all costs - into the concrete space of the *requirements inherent in the natural order*, the ontology (mode of being) of natural systems (Martins Jr., 2000).

From this perspective, space in its geological and ecosystemic dimensions must be rescued as a "privileged way of thinking and acting" (Santos, 1988), in contrast to epistemologies that have been constructed and reproduced detached from geographical reality (Pádua, 2009, p. 127).

Applied to environmental law, these premises open the creation, interpretation and application of natural and cultural heritage laws to the participation of scientific and experiential knowledge about the *fundamental qualities of the* different types of environment. It is precisely in this sense that legal knowledge - which by definition is expressed and condensed in a deontology - must re-establish its connection with *nature*, that is, with the ontological dimension of reality (Martins Jr., 2000).

In this sense, studies on geology (composition, structure, physical

---

[2] Examples of this perspective are the work of Tabarelli and Gascon (2005) and Metzger (2010).

characteristics, evolution of the processes that shape the earth), phytogeography (relationships between plants and environments), phytosociology (relationships between plants in an environment) and many others should be included in the participatory construction of legal instruments and public policies related to the natural environment.

As far as the protection of forests and other forms of vegetation is concerned, their conservation has already been the subject of extensive scientific research. However, the discoveries and advances resulting from this research have had little or no impact on the formulation and application of environmental management legislation. In general, in the creation, interpretation, application or revision of environmental legislation, little or no attention is paid to the ecosystem conditions that must be considered *a priori* for the continuity or restoration of the "good functioning" of natural systems.

In Brazil, this became clear during the recent discussions on the revision of forestry legislation, which was characterised by a clear and deliberate asynchrony between the text of the new law and the scientific proposals of the Brazilian Society for the Promotion of Science (SBPC) and the Brazilian Academy of Sciences (ABC).[4]

In this way, environmental legislation is constructed in which many legal provisions and definitions exist "only nominally" (Steigleder, 2004, p. 23) or "on paper" because they neither reflect reality nor take into account the "way of life" and "needs" of the environment.[5] Ost (1997, p. 111) sees this problem within the larger problematic framework of how the processes of creating and applying

---

[3] The wording of Federal Law 12.651 of 25 May 2012, which introduced the "new" Brazilian Forest Code that establishes the repeal of Law 4.771 of 15 September 1965, is the result of a political-legislative process in which scientific recommendations for the ethical and technical improvement of Brazilian forest legislation were not taken into account. These recommendations include those in the two editions of the volume "The Forest Code and Science: Contributions to the Dialogue" (SBPC; ABC, 2012).

[4] Against this background, Steigleder (2004, p. 23) states that: "The lack of commitment of legal concepts to truth and adequacy to reality, which underlies modern legal thinking and the formation of its institutes, makes the dialogue between ecology and law conflictual, because the increasing environmental burdens that go beyond local dimensions and individual injuries, such as diffuse pollution, acid rain, ozone hole, phenomena that are typical of a risk society, call for legal solutions other than those imposed by rules regulating individual legal relationships."

environmental regulations operate, noting that:

> In order to draw the boundaries of what is permitted and what is prohibited, to define responsibilities, to identify the parties involved, to determine the areas of application of rules in time and space, law has a habit of using definitions with clear contours, stable criteria and intangible boundaries. Ecology calls for all-encompassing concepts and evolving conditions; law responds with fixed criteria and categories that segment reality. Ecology speaks in terms of ecosystems and biospheres, law responds in terms of boundaries and limits; the one constantly develops its natural cycles, sometimes extremely long, while the other imposes the short rhythm of human predictions. And here lies the dilemma: either environmental law is the work of lawyers and cannot make sense of an extremely complex and variable situation; or the norm is written by the expert and the lawyer disowns this illegitimate child, this "engineering law", full of numbers and uncertain definitions, accompanied by endless and constantly revised lists.

The restrictive disciplinary approaches prevalent in science impede a dialogue of knowledge to create legal norms that are consistent with the promotion of social and environmental justice and the preservation of natural systems. For this reason, environmental protection is hindered by the disciplinary isolation to which the prevailing scientific culture subjects the studies dedicated to these systems. The fragmentation of knowledge that derives from the architecture of modern science ultimately contributes to the collapse of order and the complete de-characterisation of natural systems.[6] With particular acuity, Leff (2002, p. 161) states that:

> By analytically fragmenting itself in order to penetrate entities, knowledge separates what is organically articulated; without knowing it, without explicit intention, it generates a negative synergy, a vicious circle of environmental destruction that knowledge no longer understands or contains. [...] A perverse movement of knowledge that, instead of advancing by overcoming ignorance in a "dialectic of enlightenment", generates its own shadows and creates a transgenic

---

[5] For Floriani (2006, p. 68), in the sciences in general, "[...] the disjunction prevails between the multiple disciplinary knowledges, increasingly represented by specialised requirements arranged in institutionalised systems of knowledge, independently of the purposes for which they are used". In relation to legal thought, Santos (1998, p. 2) notes that the separation between the production of law and nature has its origins in the very conception of the modern state, since the social contract on which it is notionally based "rests on criteria of inclusion, which are therefore also criteria of exclusion. [The first is that the social contract includes only individuals and their associations. Nature is thus excluded from the contract, and it is significant in this respect that what is before or outside the contract is called the state of nature. The only nature that matters is human nature, and even that can only be tamed by the laws of the state and the rules of coexistence in civil society. All other nature is either a threat or a resource".

object that is no longer recognisable in the knowledge of the sciences.

This short essay-style book aims to critically analyse Brazilian forest legislation in light of the need to protect connectivity and phytoecological diversity in the Cerrado biome. To this end, the following studies are developed as specific objectives: (1) general characterisation of the biological and socioeconomic contextual aspects of the biome; (2) critical evaluation of the Constitution of the Federative Republic of Brazil and the aforementioned legislation in terms of their capacity to protect the phytoecological richness of the Cerrado; (3) Presentation of contributions to the development of analyses and research that, inspired by the idea of transdisciplinarity, lead to the resumption of a fruitful dialogue between law and environmental sciences, with the aim of formulating, interpreting and applying legal norms in a way that is compatible with the characteristics of natural systems in general and the Cerrado biome in particular.

Chapter 2

THE CERRADO BIOME: FROM BIOLOGICAL WEALTH TO

A

HIDDEN TRAGEDY

2.1 General phytophysiognomic aspects and biodiversity

According to Pádua (2009), the Bavarian traveller Carl Friedrich Philipp von Martius (1958) made one of the "most appealing" proposals for the typological systematisation of Brazilian nature. In his manuscript "A fisionomia do Reino Vegetal no Brasil" (The Physiognomy of the Plant Kingdom in Brazil) from 1824, the scholar was inspired by Greek mythology to divide the ecological diversity of the Brazilian territory into five large kingdoms symbolised by Greek gods or nymphs. Martius reserved the symbolic name "Kingdom of the Oreades" for the inner hinterland of the Cerrado. According to Greek mythology, "Oreades" were nymphs, symbols of the creative and fertile grace of nature, who ruled the fields in the service of the goddess *Diana* and inhabited and protected the mountains, caves and grottos. They were mortal, but very long-lived. They had the power to heal, prophesy and nourish.

[2]This *kingdom*, called Cerrado in the current scientific literature (Quintela, 2010), occupies about 24 % of the Brazilian territory (2 036 488 km), making it the second largest floristic province in the country (Aquino, 2006). The Cerrado borders other Brazilian biomes (Amazonas, Caatinga, Atlantic Forest and Pantanal), shares transitional areas with them and maintains gene flows. The springs and aquifers of the country's main rivers and river basins are also located in this biome (Aquino, 2006; MMA, 2010).

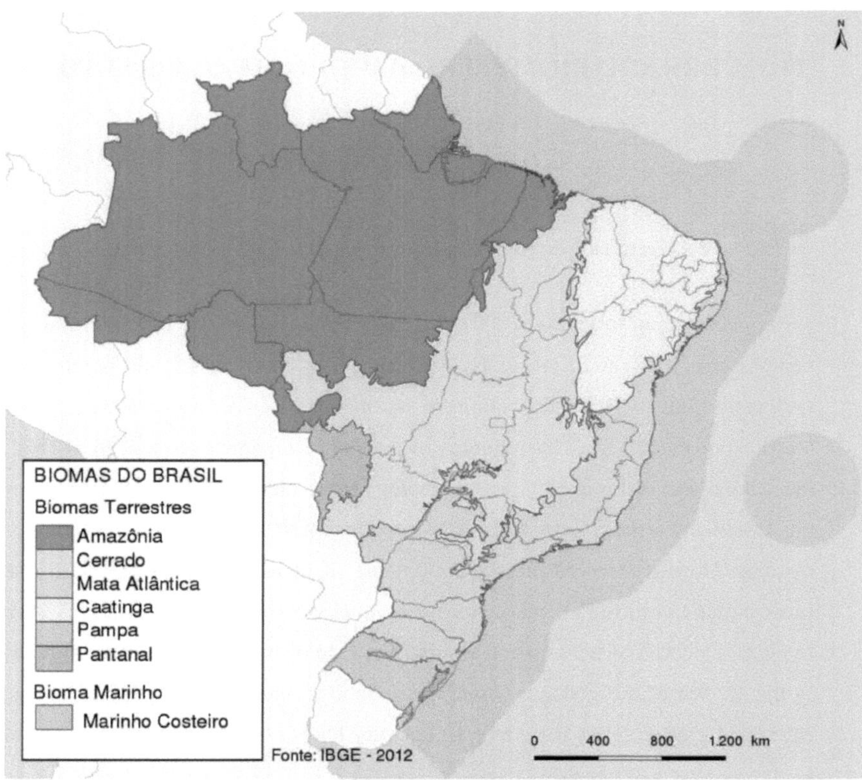

Figure 1 - Map of the Brazilian biomes (Source: IBGE, 2012).

From a phytophysiognomic point of view (vegetation form), the term "cerrado" is polysemic, as it has three different technical meanings (Walter, 2006). The first, broader meaning refers to the phytogeographic province (as a whole) that predominates in central Brazil. In this sense, the word refers to the "mosaic" of ecosystems (savannas, forests, fields and gallery forests) that exist in this province (Klink; Machado, 2005). The second term, cerrado in the broader sense, refers to the savannah and grassland formations of the biome, which extend from the cerradão with its dense tree structure to the Campo Limpo. "Under this term" - according to Walter (2006, p. 37) - "only one [typical] forest formation is included, the Cerradão". For the ecologist, the last meaning, cerrado in the narrower sense, is the one that best identifies and represents the biome. It approximates the usual concept of savannah and refers to a "tropical formation dominated by grasses and containing a greater or lesser proportion of open woody vegetation and associated trees" (Collinson, 1988 *apud* Walter, 2006, p. 37).

As Henriques (2005) explains, the Cerrado biome consists of a heterogeneous mosaic of plant physiognomies. In this mosaic there is a height-density gradient, with grassland physiognomies at one end and forest physiognomies at the other (Figure 2). This gradient forms a phytophysiognomic *continuum*, as there are no clear boundaries between one plant physiognomy and another and intermediate forms often exist in between (Henriques, 2005).

In the Cerrado there are other types of vegetation, such as wet meadows and gallery forests, which are structurally and physiognomically similar to those of the gradient mentioned above, but differ due to their particular floristic composition and soil factors (Henriques, 2005).

Figure 2 - Illustration of the vegetation types in the Cerrado biome. Author: José Felipe Ribeiro. Available at: <http://bit.ly/1OkoN8q>. Retrieved on: 08 June 2017.

The typical savannah vegetation takes up most of the biome. According to Eiten (1972, 1977, 1978), quoted by Walter (2006), 80 to 90 per cent of central Brazil. These data are relevant given the - sad - observation that in many parts of the world, especially in Brazil, environments without dense tree cover do not raise the awareness of social and institutional actors as much as those that common sense recognises as forests. The scientist Frederico Carlos Hoehne (1930) from Minas Gerais, quoted by Franco and Drummond (2005, p. 6), already argued in a 1930 work that not only forests but also Brazil's other phytophysiognomies are worthy of recognition. Alluding to the Cerrado, the scholar, who was one of the first to conduct systematic, comprehensive and long-term research into the country's flora, lamented that "the rough fields [...] have never deserved our attention. The flames have devoured them without us ever realising that they are admirable, rich and gifted".

Although it is home to a remarkable biological wealth, the Cerrado, especially in its savannah-like characteristics, is not recognised as important by many social and institutional actors. Maciel (2008) identifies a "discourse of inferiority of the Cerrado". In the national context, the forest physiognomies of the Amazon and the remnants of the Atlantic rainforest are receiving more attention from environmental movements and government agencies (Walter, 2006). Even in geography and science textbooks, interest in the country's second largest biome is neglected, according to a recent study (Bizerril, 2008). The savannahs that predominate in the Cerrado are seen as less important vegetation. "This is a mistake!" - Walter (2006, p. 35) denounces. "Natural savannas are a biological fact and are important because they cover large parts of the planet and can be just as rich as the richest tropical forests, as is the case in the Brazilian Cerrado."

Along with the Atlantic Rainforest, the Cerrado is one of the biomes on the list of *hotspots* for biodiversity conservation (Conservation International, 1999). In fact, it is a strategic area for biodiversity conservation, as it has a high biodiversity per square metre, but is seriously threatened by human activities.

In terms of floristic composition, the number of vascular plants in the Cerrado is higher than in most other regions of the world. Herbaceous plants, shrubs, trees and lianas total around 10,000 species, and studies predict that this number could double (Novaes, 2008). In addition, 44 % of the flora is endemic, i.e. its distribution is restricted to the biome area (Klink; Machado, 2005). The fauna is equally rich, with 159 species of mammals, 23 of which are endemic. The number of catalogued bird species amounts to 837 (29 endemic). The Cerrado is also home to around 180 reptile species (20 endemic) and 113 amphibian species (32 endemic) (Aquino, 2006).

All these indicators make the Cerrado one of the richest tropical natural systems on Earth, emphasising its uniqueness and "biological dignity" (Alvarenga, 2013). A dignity sufficient to justify its maximum protection, in the light of a reason that, taking into account the diachronic breadth of the fundamental right to the environment, explicitly enshrined in the Constitution of the Federative Republic of Brazil (CRFB), gives the biotope the value of a true natural heritage. A heritage that, after a slow geological and ecosystemic evolution, has translated into landscape ensembles with a long and complex landscape development.[7]

2.2 Socio-economic utilisation scenario

Although the international scientific community recognises the Cerrado as a *hotspot* for the conservation of biodiversity, deforestation is progressing *rapidly*, especially for the expansion of agricultural frontiers. So fast, in fact, that some estimates predict the disappearance of the biome before the beginning of the second half of the 21st century (Machado *et al.*, 2004). This is a grim foretaste of a scenario that could occur as a result of a silent catastrophe and, at an institutional level, a dilemma that underlies the definition of policies and legal norms related to the natural resources of the Cerrado. While on the one hand social movements and researchers are calling for the protection and restoration of the region, on the other hand the so-called "rural faction" is demanding the use of hundreds of thousands of additional hectares for the expansion of agribusiness and other economic activities, such as the creation of new industrial parks and the construction of new

---

[6] As Ab'Sáber (2003, p. 10) states, "More than territorial spaces, peoples have inherited landscapes and ecologies for which they are or should certainly be responsible. From the highest levels of government and administration to the most ordinary citizen, everyone has a permanent shared responsibility for the non-predatory use of this unique heritage that is the Earth's landscape".

industrial centres.For example, the creation of new industrial parks and the parcelling out of land for urban purposes.[8] As Ab'Sáber (2003, p. 24) notes, in Brazil in general there is still "an atavistic attachment to the vast forested 'hinterland' that made life difficult for the first settlers". In 2002, a study based on Modis satellite imagery concluded that 55% of the biome's vegetation had already been displaced or heavily modified by human intervention. This percentage corresponds to almost three times the amount of vegetation that had been removed from the Amazon region by that year (Machado *et al.*, 2004). Annual deforestation rates are also higher in the Cerrado. [2]Between 1970 and 1975, average deforestation in this biotope reached a staggering 40,000 km/year, 1.8 times the rate of deforestation in the Amazon in the period 1978-1988. The current extent of forest destruction in the Cerrado is no less grim. [2]While the attention of the Ministry of the Environment (MMA) and the international community is focused on combating deforestation in the Amazon, the Cerrado is losing an average of 21.26 kilometres of vegetation per year, according to recent data from the Ministry of the Environment (Salomon, 2009). Environments that were once rich in biological, geological and cultural diversity are rapidly being transformed into homogeneous areas for extensive cereal cultivation. Between 2002 and 2008, the biotope suffered vegetation losses equivalent to half the area of the state of São Paulo. This deforestation rate is more than double the estimated deforestation rate

---

[7] It is argued, among other things, that the natural conditions of the Cerrado favour agricultural expansion. "Because the Cerrado is located on a topography favourable to agriculture," notes Mantovani (2003, p. 393), "it is a biome that has been devastated throughout the country, and because it is rich in grass species, it is used as natural pasture for extensive livestock farming, stimulated by the felling and thinning of shrubs and trees, making nutrients available in the atmosphere and in the soil through burning."

According to the Brazilian Society for Agricultural Research (Embrapa), less than 5 % of the total area of the biome has residual vegetation of more than 2,000 hectares (Novaes, 2008). Figure 3 shows the cumulative deforestation up to 2009 at the biome level. Visual interpretation clearly shows the loss of continuity of vegetation along the Cerrado.

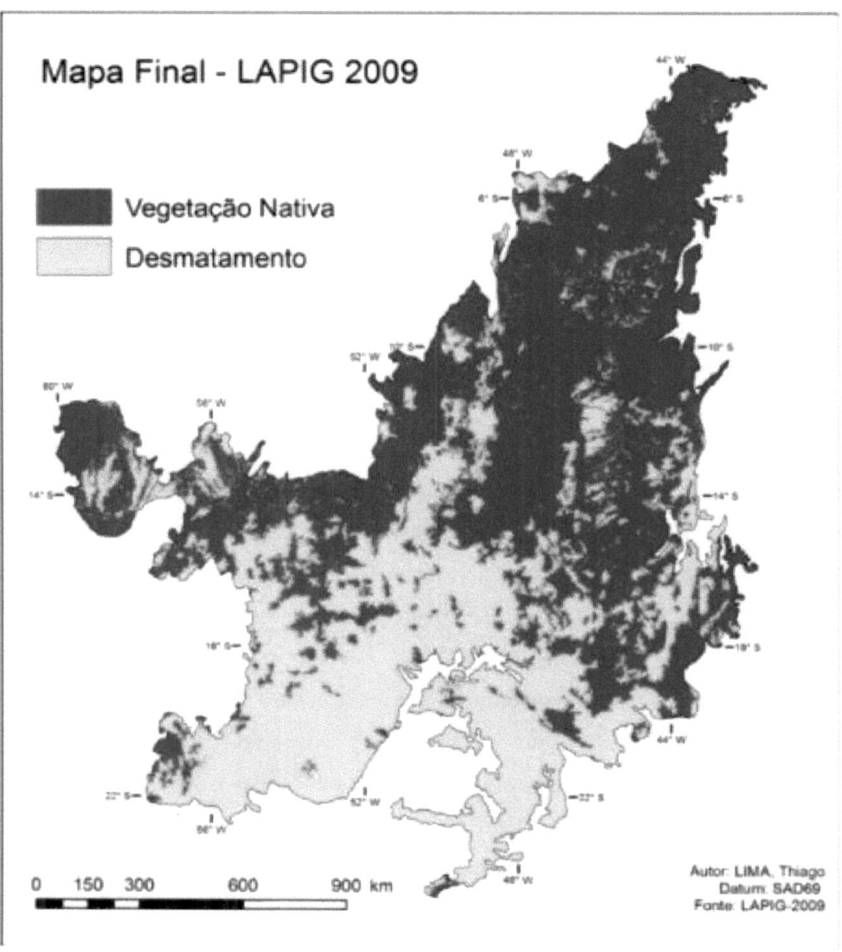

Figure 3 - Siad/Modis map of deforestation in the Cerrado biome. Source: Probio; Lapig/Siad-Cerrado.

From a socio-ecological perspective, various traditional peoples and communities in the Cerrado, such as the Veredeiros, Geraizeiros, Vazanteiros and Barranqueiros, have developed sustainable forms of management of the biome's ecosystems over several generations. Currently, the landscapes of many of these communities are threatened by the violence of land grabbing and agribusiness, which has led these peoples to fight for the creation of Conservation Units (CUs), especially for sustainable use, as a strategy for remaining in their territory. However, the land converted into protected areas is not enough to contain the environmental degradation in the Cerrado (Aquino, 2006) and allow the reproduction of sustainable lifestyles at the scale of the biome.

All these facts form a sad and contrasting scenario to the scientific recommendation that authorities at national, regional and local levels should aim for "avoided deforestation" (Lavratti; Prestes, 2009) for the Cerrado, at least until an integrated and strategic planning for the utilisation of the biome's natural resources is developed (Machado *et al.*, 2004). Until this happens, the changes in the region will continue and will be accompanied by numerous negative socio-ecological impacts, such as habitat fragmentation, biodiversity decline, exotic species invasion, soil erosion and compaction, aquifer pollution, ecosystem degradation, alteration of combustion regimes, carbon cycle imbalance, regional climate change, loss of nutrients and erosion of socio-ecological diversity (Klink; Machado, 2005).

In addition, the susceptibility of large parts of the Cerrado to desertification due to anthropogenic factors (deforestation, excessive grazing) should be emphasised, especially in some regions of the states of Ceará and Piauí, where this process is already well advanced (Mantovani, 2003, p. 397). All these changes taken together could have a negative impact on other natural areas of the country. As Sawyer (2008, p. 4) warns:

> Water, biodiversity and climate are interdependent. Water depends on the vegetation cover, i.e. the flora, whose reproduction in turn depends on the fauna, which pollinates the flowers and spreads seeds and spores. Flora and fauna are dependent on rainfall and the flow of paths, streams and rivers. If one of these links fails, the vital chain is broken and the

entire ecosystem can collapse. Diversity is an important factor in adapting to climate change. If Brazil's central ecosystems collapse, the other ecosystems will also suffer.

Given this situation, one of the main challenges for the conservation of the Cerrado is to emphasise the social, scientific, medical, cultural, etc. importance of the biological, geographical and socio-economic diversity of the biome. The discussions that form the basis for the development of legislation and management plans for the natural province in question. The discussions that form the basis for the development of legislation and management plans for the natural province in question must take into account the body of scientific information "both on species and habitats and on the functioning of ecosystems, since changes in the landscape have implications for the burning regime, hydrology, carbon cycle and stock, and possibly the climate" (Klink; Machado, 2005, p. 152).[9]

2.3 The "silence" of the Brazilian constitution and the devastation of the Cerrado

The high level of destruction and fragmentation of the Cerrado is partly due to the discriminatory way in which some key provisions of the Brazilian legal system treat the country's existing biomes. The fact that the great Brazilian savannah is not one of the natural regions that the CRFB considers to be "national heritage" contributes to the neglect with which social, political and economic actors treat it, according to specialised literature. The Cerrado, the Caatinga and the Pampa are considered by many to be "poor cousins" among the Brazilian biomes.

The "silence" of the CRFB in relation to the Cerrado has negative consequences for the treatment of the biome at the infra-constitutional level. Take the example of the Legal Reserve (RL): while rural landowners in the Amazon

---

[8] Henriques (2003, p. 36) also points out that "the environmental scenario - in the Cerrado and the neighbouring regions - that will affect the use of natural resources, such as water, and the quality of life of Brazilians at the end of the first half of the 21st century is a very different one. Only in this way will it be possible to avoid or minimise the negative effects of the destruction and degradation of the Cerrado".

region are obliged to maintain 80 % of the vegetation cover on their land, the proportion of the floral reserve in the Cerrado is no more than 20 %.

The "silence" of the constitutional text on the Cerrado, which is the result of prejudices that minimise the biological importance of the forms of vegetation found in this biome due to ignorance, is therefore one of the factors leading to severe deforestation and fragmentation of the ecosystems of the great Brazilian savannah.

In addition, *forest legislation* does not consider the conservation of vegetation connectivity at the river basin level as mandatory, even if only as a technical guideline (see *below*). In general, federal and state laws do not contain provisions that oblige or incentivise rural landowners or public authorities to maintain the *continuum of the* different phytophysiognomies of the Cerrado.

The lack of constitutional protection of the Cerrado and the inadequate regulation of the conservation of the plant-ecological network have contributed to the devastation and fragmentation of the biome, with the result that it is in danger of disappearing completely in the next twenty to thirty years.

2.4 Forestry legislation and phytoecological protection

Maintaining the connectivity of vegetation is a necessary condition for the "proper functioning" of the natural system in which it is distributed, and some of the negative effects of the loss of this connectivity are well known. These include: (1) the reduction and modification of habitats favourable to species life and reproduction; (2) aggregation effects, such as. imbalance of populations and increased competition for resources; (3) isolation and disruption of the spatial dynamics of species; (4) extinction of species; (5) introduction of alien species; (6) edge effects and increased vulnerability of remaining vegetation (Rambaldi; Oliveira, 2005; Tabarelli; Gascon, 2005; Kettunen *et al*, 2007).

These negative effects on the conservation of the environment, especially the forests, include the edge effects described didactically by Fernandez (2011, p. 152 ff.). In the case of the Cerrado, these effects mainly affect the forest (cerradão) and shrub (cerrado senso estrito) physiognomy of the biome.

However, Brazilian forestry legislation does not yet contain regulations that oblige social actors, whether public or private, to maintain the spatial continuity of vegetation, especially in the Cerrado biome. This conclusion is based on the following observations:

(a) Taking the river basin as a territorial unit of analysis, it is evident that licences and permits for logging are often granted without consideration of the need to maintain connectivity between vegetation remnants on rural land;

(b) The size and location of the permanent protected areas (APP) and the RL are based on a quantitative spatialisation logic typical of Euclidean geometry. The size and allocation of these Special Protection Areas are the result of fixed percentages established in advance by law. In most cases, the specific characteristics of each environment and the complex chains of ecosystem interrelationships that exist at the river basin level are not taken into account;

(c) Although the water catchment area is included in the laws on agricultural policy and the National Water Resources Policy as a territorial unit for environmental analysis and environmental management, in practice it is not taken into account for these purposes. Thus, the delineation of water catchment areas is mostly limited to the boundaries of land parcels. There is a lack of a systemic, comprehensive view of the entire catchment area to determine the most suitable *locations* for these Etep. Ultimately, forest legislation and associated management practices contribute to the creation of environments with fragmented, i.e. spatially disjointed, vegetation.

These observations are confirmed by a historical-geographical analysis of the colonisation of two representative parts of the biome. The first consists of an enrichment zone of the Entre Ribeiros River sub-basin and the São Marcos River sub-basin in the Paracatu region, northwest of Minas Gerais, on the border with the Federal District. Comparing the vegetation cover in this area between 1964, the year before the previous forestry law (Law 4.771/1965) was passed, and 2005, it can be seen that the "mosaic" of native forests formed by the different phytophysiognomies of the Cerrado has given way to human activities, especially agribusiness. As a result of the aforementioned "prejudices" against the Cerrado, which particularly affect the savannah characteristics of the biome, the areas of grassland, Cerrado grassland and Cerrado senso estrito have decreased significantly during the period in question (see Table 1).

Accordingly, there have been no consistent political and administrative efforts to establish ecological corridors to connect the remaining vegetation - an omission that in practice jeopardises the conservation of natural systems in the region. The figures below are a cartographic representation of the loss of vegetation connectivity in the region in question.

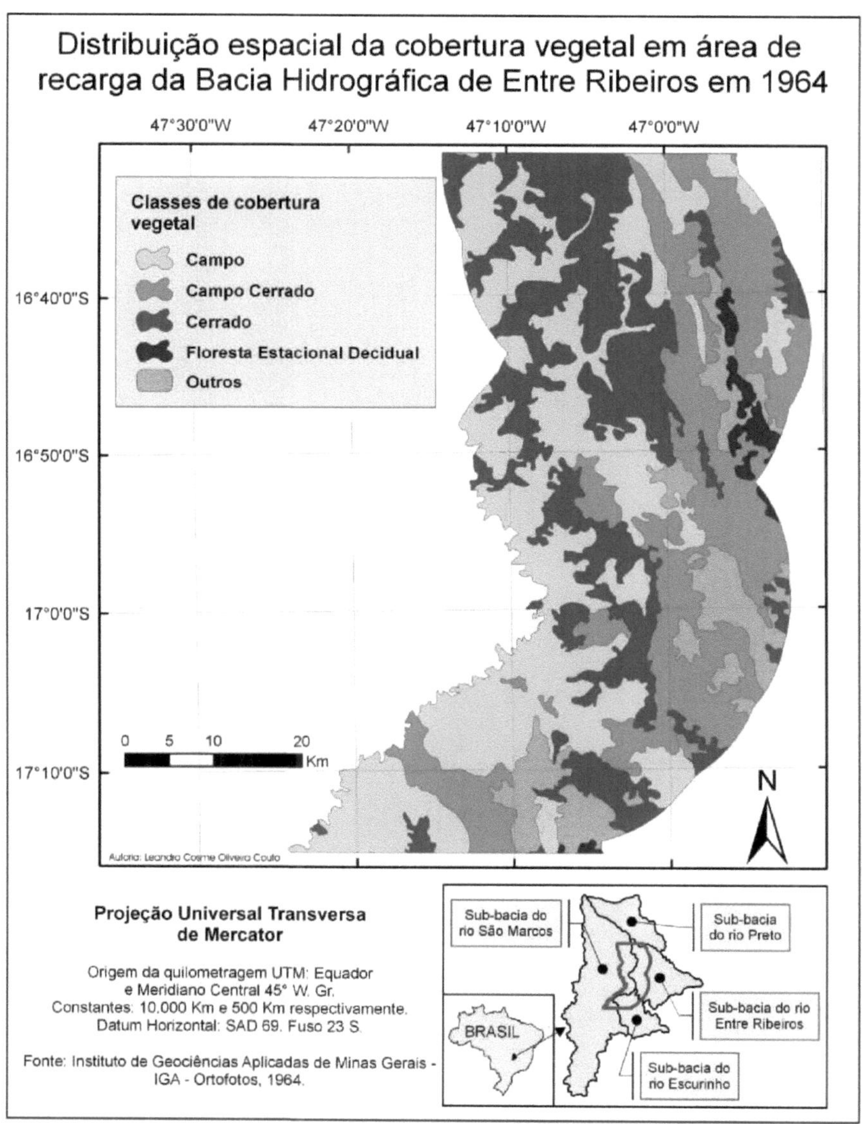

Figure 4 - Map of the particular distribution of vegetation cover in the
Entre Ribeiros catchment area
in
1964
.

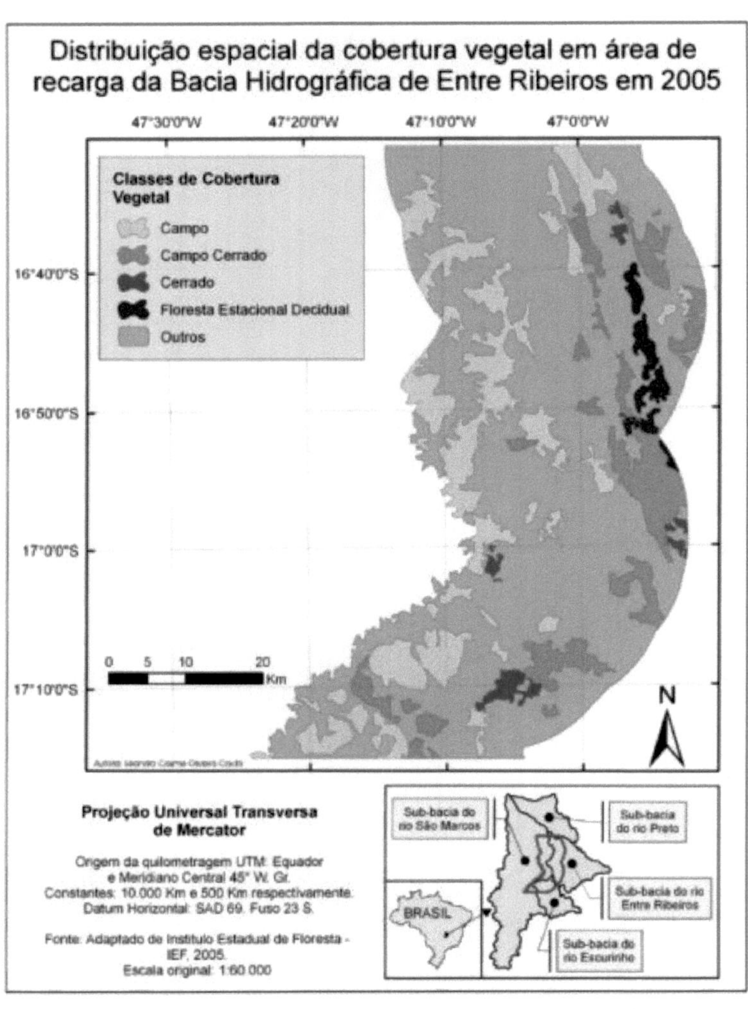

Figure 5 - Map of the particular distribution of vegetation cover in the
Entre Ribeiros catchment area
in
2005
.

Table 1 - Quantification of temporal variations in vegetation cover in the Entre Ribeiros catchment recharge area in the period 1964-2005.

CLASSES OF VEGETATION COVER	Area in 1964		Area in 2005		Change between 1964 and 2005		
	km²	% / area	km²	% / area	km²	% / area	% related to the class
Country life (altitude, clean and dirty)	1.010,9	36,00	491,0	17,49	-519,9	-18,51	-51,43
Cerrado field	745,5	26,55	272,4	9,70	-473,1	-16,85	-63,46
Cerrado	745,1	26,54	37,6	1,34	-707,6	-25,20	-94,96
Seasonal forest Decidual	49,4	1,76	49,3	1,76	-0,1	0,00	-0,14
Other (environments changed; Agriculture, Livestock farming, urbanisation, exposed soil, etc.)	257,1	9,15	1.957,7	69,72	1700,6	60,56	661,60
Total native vegetation	2.551,0	90,8	850,3	30,28	-1.700,6	-60,56	-66,67
Total area in comparison	2.808,0	100,00	2.808,0		100,00		-

As can be seen, the landscape has changed drastically in the period 1964-2005: The area used by humans has increased by 661.60 % during this period, corresponding to 60.56 % of the area previously occupied by the diverse and biologically rich plant species of the Cerrado biome, totalling 69.72 %. The plant subspecies with savannah physiognomy have experienced an area reduction of more than 50 %. In 1964, the Cerrado and the Campo Cerrado occupied almost equal areas; however, the Cerrado showed a greater reduction in area (94.96 %), while the Campo Cerrado showed a reduction of 63.46 %, corresponding to 1.34 % and 9.70 % of the area in 2005, respectively. The Campo Cerrado also showed a large decrease in area, 51.43%, while the Estacional Decidual Forest - with a typical forest physiognomy - showed a minimal decrease of 0.14%, which is tiny in spatial terms.Considering the interrelationships between the different ecosystem elements, it can be said that the significant decrease in forests and the consequent loss of vegetation connectivity places the region in a situation of environmental vulnerability. This is because these anthropogenic processes affect the survival conditions of fauna and flora, increase the susceptibility of the soil to erosion and desertification, alter the water cycle and have other negative effects.

Further studies in the north-western region of Minas Gerais, on the border between the states of Minas Gerais, Goiás and the Federal District, came to no less worrying results. The rapid destruction of the spatial continuity of vegetation in the Cerrado shows how accurate the scientific predictions are that the sustainability of the biome as an ecological *continuum* is seriously jeopardised by the spatial encroachment of human activities. The loss of the phytophysiognomy of the great Brazilian savannah was observed in a strip of almost 600 kilometres in length, extending from the region of the municipality of Vazante, Minas Gerais, to the Federal District, at the upper end of the watershed of several river basins. Three years were used as a reference for these studies: 1964 (one year before the enactment of Law 4.771, the forestry law repealed in 2012 by Law 12.651 of 2012), 1989 and 2005.

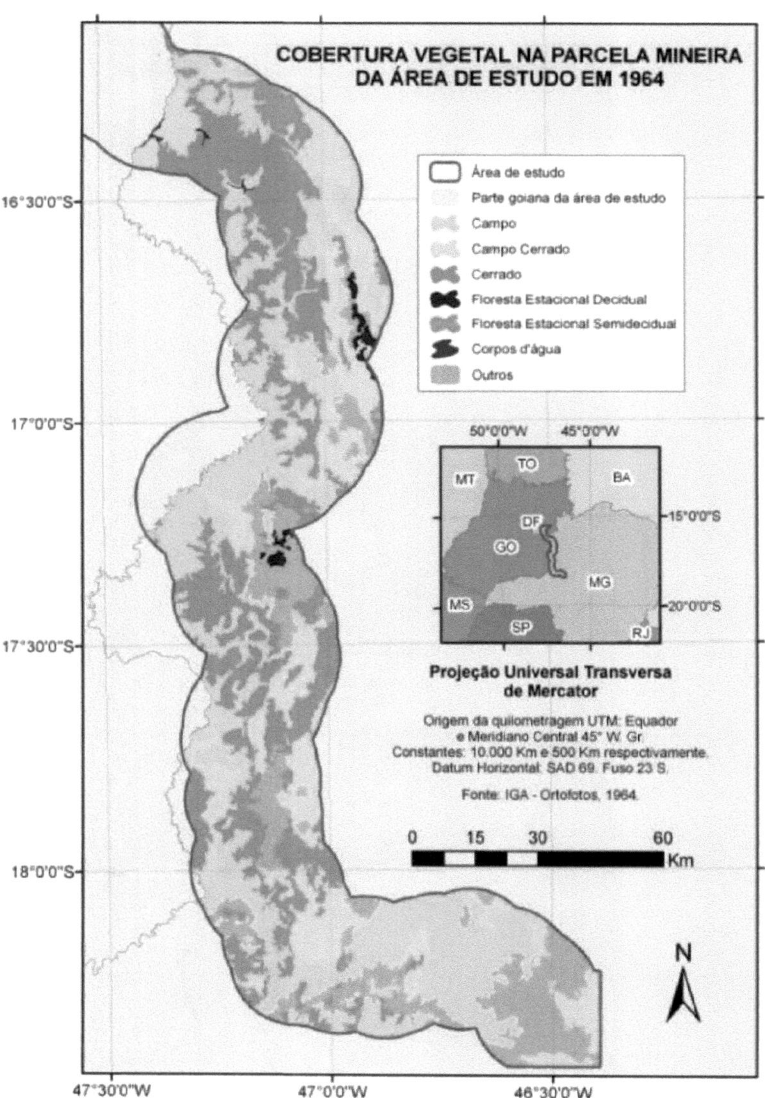

Illustration 6 - Hilltop axis between the catchment areas of Paracatu/MG, São Marcos/GO, Alto Paranaíba/MG and the
São Mateus/GO in 1964 (Martins Jr. *et al.*, 2009) in 1964.

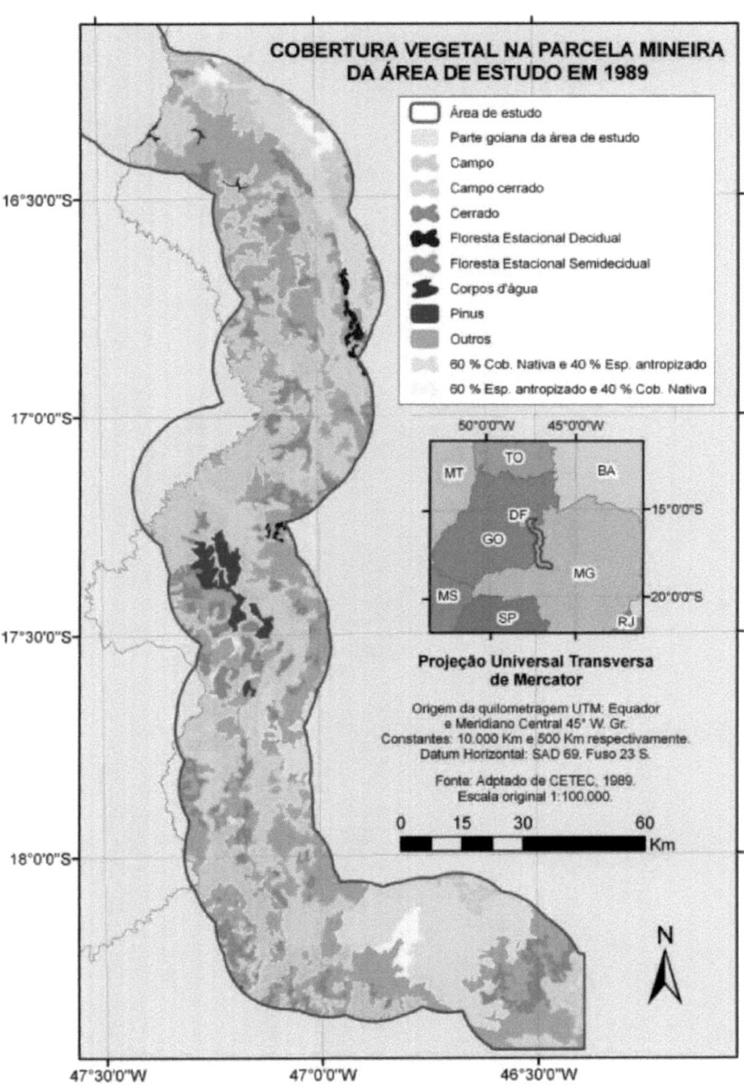

Illustration 7 - Hilltop axis between the catchment areas Paracatu/MG, São Marcos/GO, Alto Paranaíba/MG and the São Mateus/GO in 1964 (Martins Jr. *et al.*, 2009) and 1989.

2005

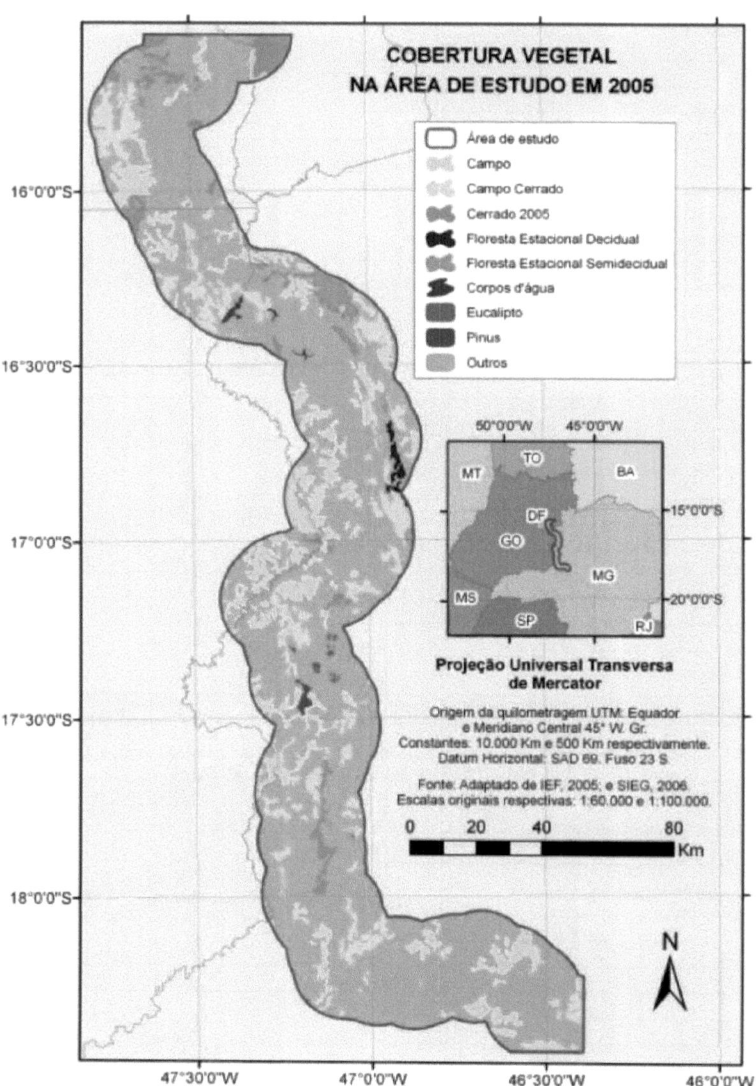

Illustration 8- Hilltop axis between the catchment areas Paracatu/MG, São Marcos/GO, Alto Paranaíba/MG and the
São Mateus/GO in 1964 (Martins Jr. *et al.*, 2009) in 2005.

Table 2 - Quantification of vegetation cover on the mountaintop axis between the Paracatu/MG, São Marcos/GO, Alto Paranaíba/MG and São Mateus/GO catchments in 1964 (Martins Jr. *et al.*, 2009) in 1964.

Vegetation cover classes / Anthropogenic space	Range		General classes of room utilisation	Range	
	Hectare	%			
Country life (altitude, clean and dirty)	311.942,0	30,84	Areas with natural vegetation	850.383,7	84,07
Cerrado landscape	247.286,6	24,45			
Cerrado	267.628,6	26,46			
Seasonal deciduous forest	7.141,5	0,71			
Deciduous seasonal forest	16.384,9	1,62			
Water surfaces	1.414,1	0,14	Water surfaces	1.414,1	0,14
Pine	0,0	0,00	Anthropised space	159.672,4	15,79
Eucalyptus	0,0	0,00			
Other (cultivated land, urban areas, pastures, exposed soils)	159.672,4	15,79			
In total	1.011.470,2	100,00	In total	1.011.470,2	100,00

Table 3 - Quantification of vegetation cover on the mountaintop axis between the Paracatu/MG, São Marcos/GO, Alto Paranaíba/MG and São Mateus/GO catchments in 1964 (Martins Jr. *et al.*, 2009) in 1989.

Vegetation cover classes / Anthropogenic space	Range		General classes of room utilisation	Range	
	Hectare	%		Hectare	%
Country life (altitude, clean and dirty)	286.640,4	28,34	Areas with natural vegetation	682.287,4	67,46
Cerrado field	286.874,7	28,36			
Cerrado	92.329,4	9,13			
Seasonal deciduous forest	5.248,4	0,52			
Deciduous seasonal forest	11.194,5	1,11			
Water surfaces	1.416,9	0,14	Water surfaces	1.416,9	00,14
Pine	16.436,1	1,62	Anthropised space	327.765,9	32,40
Eucalyptus	0,0	0,00			
Other (cultivated land, urban areas, pastures and open ground)	311.329,7	30,78			
In total	1.011.470,2	100,00	In total	1.011.470,2	100,00

Table 4 - Quantification of vegetation cover on the mountaintop axis between the Paracatu/MG, São Marcos/GO, Alto Paranaíba/MG and São Mateus/GO catchments in 1964 (Martins Jr. *et al.*, 2009) in 2005.

Vegetation cover classes / Anthropogenic space	Range		General classes of room utilisation	Range	
	Hectare	%		Hectare	%
Country life (altitude, clean and dirty)	115.173,0	11,39	Areas with natural vegetation	231.647,7	22,90
Cerrado landscape	81.909,8	8,10			
Cerrado	13.459,4	1,33			
Seasonal deciduous forest	4.930,7	0,49			
Deciduous seasonal forest	16.174,8	1,60			
Water surfaces	1.419,2	0,14	Water surfaces	1.419,2	0,14
Pine	2.611,1	0,26	Anthropised space	778.403,2	76,96
Eucalyptus	1.448,5	0,14			
Other (cultivated land, urban areas, pastures and open ground)	774.343,6	76,56			
In total	1.011.470,2	100,00	In total	1.011.470,2	100,00

In 1964, for example, the proportion of high-altitude grassland was 30.84 %

of the area; in 2005, it fell to 11.39 %. Looking at the total area, it can be seen that the area deforested for agricultural purposes in particular totalled 15.79% in that year. In 1989, this percentage more than doubled to 32.4 %. In 2005, the degree of anthropogenic change in the region was 76.96 %, which is an excellent development. Apart from a few exceptions, mostly confined to the boundaries of a few protected areas, *this pattern of environmental change has spread across the entire biome*. Figures 9 and 10 illustrate the types of interventions that have taken place in the area representative of the processes discussed here. The spatial continuity between the phytophysiognomies of the Cerrado biome has been disrupted, particularly by the conversion of land for intensive agriculture.

Figure 9 - Agricultural field in a typical mountain landscape in the Paracatu Valley; the colonisation was due to the simple agricultural activity in largely flat areas. Photo: Martins Jr. (2005).

Figure 10 - Water consumption with inefficient irrigation and high wastage due to excessive evaporation in the Paracatu Valley. Photo: Martins Jr. (2005).

In this problematic context, it is an urgent task to review the concept of "conservation". The reconstruction of this concept must be based on positive forms of development in which the ontological characteristics of natural systems are concretely taken into account, which means respecting the natural destination of spaces and not creating situations of irreversibility.

This conceptualisation is based on the possibility of *cooperation* between human action and the specific characteristics of natural systems in the light of the postulates of the ethics of care (Boff, 2009). In the field of juridical thought, this is a development based on the "consideration of natural logic 'in itself'" (Ost, 1995, p. 112) for the design and implementation of norms, policies and concrete initiatives to protect the environment.

Chapter 3

LAW, SCIENCE AND THE CERRADO BIOME

The degradation and fragmentation of natural systems in Brazil has continued to intensify since the colonial era. The logic behind these processes is constantly reproduced in the country's social and institutional practices. The natural environment is generally seen as an "obstacle" to the free development of productive forces, and government and market actors still face the "old dilemma" that sacrificing primary vegetation is unavoidable for better economic utilisation of space (Ab'Sáber, 2003).

Following this logic, environmental destruction can be observed in all regions of Brazil. In the Cerrado in particular, it is progressing by leaps and bounds: the region's ecosystems are being ostentatiously removed to make way for agriculture and livestock farming. The cartographic information in this paper shows the accelerated displacement of the biome's phytophysiognomies by the introduction of human activities over a period of forty years.

However, the sciences can help to ensure that the environment, especially in the biomes, is used in a truly sustainable way, beyond concepts that are distant from the landscape and rhetoric that is characterised by insidious interests.

To this end, the design of public policy and environmental norms must be open to *dialogue* with other forms of conservation knowledge. Environmental normativity must be developed through attentive listening to forms of knowledge expressed in 'environmental needs', whether they derive from science or from the human experience of the place. This is not, of course, to deny the scientific autonomy of law vis-à-vis other sciences and knowledges, but to recognise that the complexity of the treatment of environmental goods must match the *complexity of legal-ecological epistemology*, which rests on the methodological and conceptual openness of the processes of normative production and reproduction to the consideration of other forms of knowledge.

Scientific knowledge about the need to maintain phytoecological connectivity between the vegetation physiognomies of the Cerrado must be integrated into the processes of creation, interpretation and application of legal norms. However, the current forest legislation, which is the result of a culture and constitution that undervalues the region, is not designed nor has it been operationalised to preserve the spatial continuity between the diverse and rich phytophysiognomies of the biome.

Forest legislation should have the conservation of Brazilian flora as a central objective, in response to the warnings of science and in accordance with the political and legal provisions of the Federal Constitution, which imposes on the authorities the duty to "maintain and restore the essential ecological processes and provide for the ecological management of species and ecosystems" and "to protect the animal and plant world, prohibiting, in the form of laws, practices that jeopardise their ecological function, cause the extinction of species or treat animals cruelly" (Art. 225, §1, I and VII).

Maintaining the spatial continuity of vegetation is a necessary condition for the "proper functioning" of the natural system in which it is distributed (Rambaldi; Oliveira, 2005; Kettunen *et al.*, 2007), and some of the effects of the loss of this connectivity are already well known to science.

However, environmental and forestry legislation does not yet contain provisions that, in recognition of this *precondition* for the conservation of geosystems, proportionately combine *incentives* (funding rights) or *obligations* (command and control logic) for public and private social actors to conserve the connectivity of vegetation forms.

As Ost (1997, p. 129) notes, environmental regulations have often been drafted out of a lack of commitment to territorial requirements, without taking into account the specific local characteristics of the environment. As for the relationship between fragmentation

notes the Belgian lawyer:

> Fragments of unspoilt nature will never constitute a viable biotope, just as a few favoured species will not be able to maintain biodiversity at a satisfactory level. What is the point of designating a wetland as a nature reserve if pollution from outside continues to disturb the balance of the environment? And what is the point of protecting this or that butterfly if the host plant of that species disappears?
>
> [...] very often the environmental standard is less the result of genuine ecological scientific requirements than of provisional concessions by interested industrial circles [...].
>
> Finally, far from defining an overall status for species and natural spaces

that guarantees their quantitative and qualitative protection, environmental law seems to be dividing spaces into countless distinct zones and subdividing resources into a multitude of special regimes, leaving them each to ever more specific transformative uses, to which a complacent legal framework is offered that definitively condemns nothing but obvious abuses (Ost, 1997, pp. 113, 129).

In this context, Delalibera *et al.* (2008) point out that although environmental and forestry legislation contributes to environmental management, it does not clearly and objectively specify or consider fundamental concepts for habitat conservation, such as: (1) the connectivity of vegetation at the watershed scale; (2) the width of ecological corridors; (3) the size of vegetation remnants for conservation effectiveness; (4) the edge effect. This highlights the importance of spatial planning strategies that prioritise the integration of protected areas and the creation of ecological corridors based on a comprehensive view of the river basin.

In addition to these strategies, there is a need for an interdisciplinary and democratic approach to the development, review, interpretation and application of environmental and forestry legislation.

Faced with this problematic and challenging situation, we are aware that the fate of the *Kingdom of the Oreades* depends to some extent on an epistemological reinvention of the law capable of opening it up to a dialogue with scientific and experiential knowledge about nature in general and the Cerrado in particular.

It is to be hoped that the law will one day be able to recognise the dignity of the biome and thus guarantee it protection that is commensurate with and appropriate to its biological and cultural value. However, most of the efforts to protect and restore the Cerrado are reserved for the concrete actions of people of good will, characterised by environmental ethics...

REFERENCES

AB'SÁBER, A.N. *The areas of nature in Brazil*: landscape potential. São Paulo: Ateliê Editorial, 2003.

ALVARENGA, L.J. *A conservação do bioma Cerrado*: o Direito ante a fragmentação de ciências e ecossistemas. São Paulo: Annablume, 2013.

ANDRADE, C.D. *Mata Atlântica*. Rio de Janeiro: AC&M Ed.; Sette Letras, 1997.

ANTONIL, A.J. Cultura e opulência no Brasil por suas drogas e minas, 1711. available at: <http://bit.ly/2shNgKM>. Accessed on: 2 Apr. 2008.

AQUINO, F.G. *Cerrado and Caatinga*: forgotten national heritage? Available at: <http://bit.ly/2s4G3fd>. Retrieved on: 10 August 2006.

BIZERRIL, M.X.A. *The Cerrado in geographical and scientific textbooks*. Available at: <http://bit.ly/2svgA1a>. Retrieved on: 29 April 2008.

BOFF, L. *St Francis of Assisi*: tenderness and strength: a reading from the poor. 12th edition. Petrópolis: Vozes, 2009.

BRANDON, K. *et al*. Brazilian conservation: challenges and opportunities. *Megadiversidade*, Belo Horizonte, No. 1, pp. 7-13, Jul. 2005.

BRAZIL, Constitution of the Republic, 5 October 1988. Available at: <http://bit.ly/VxoKkD>. Retrieved on: 21 May 2012.

BRAZIL, Law 4.771 of 15 September 1965, available at: <http://www.planalto.gov.br/ccivil 03/leis/L4771 .htm>. Retrieved on: 6 August 2017.

BRAZIL. Law 8.171 of 17 January 1991, available at: <http ://bit.ly/2tfoGru>. Retrieved on: 21 May 2012.

BRAZIL, Law 9.433 of 8 January 1997, available at: <http://bit.ly/1iC6rDg>. Retrieved on: 21 May 2012.

BRAZIL. Law 12.651 of 25 May 2012. available at: <http://bit.ly/1zecCID>. Retrieved on: 09 June 2017.

COLLINSON, A.S. *Introduction to World Vegetation.* 2nd edition. London: Unwin Hyman Ltd, 1988.

CONSERVATION INTERNATIONAL (Brazil). *Hotspots*: the most biologically rich and threatened regions of the world, 1999. available at: <http://bitly/2rp7gHy> Accessed on: 18 July 2007.

DELALIBERA, H. C. *et al.* Legal reserve allocation in rural properties: from cartesian to holistic. *Brazilian Journal of Agricultural and Environmental Engineering*, No. 12, pp. 286-292, 2008.

DUSSEL, E. *1492*: Die Verheimlichung des Anderen: der Ursprung des Mythos der Moderne: Frankfurter Vorlesungen. Translated by J.A. Clasen. Petrópolis: Vozes, 1993.

EITEN, G. A sketch of the vegetation of central Brazil. *In*: Latin American Botanical Congress, National Botanical Congress, *Anais...*, pp. 1-37, 1978.

EITEN, G. Brazilian savannas. *In*: HUNTLEY, B.J.; WALKER, B.H. (eds.). *Ecology of tropical savannas*. Berlin: Springer Verlag, pp. 24-27, 1982.

EITEN, G. Delimitation of the cerrado concept. *Arquivos do Jardim Botânico*, Rio de Janeiro, n. 21, pp. 125-134, 1977.

EITEN, G. The Cerrado vegetation of Brazil. *The Botanical Review*, New York, No. 38, pp. 201-341, 1972.

FEYERABEND, P. *Farewell to reason*. Coimbra: Edições 70, 1991.

FRANCO, J.L.A.; DRUMMOND, J.A. Frederico Carlos Hoehne: the importance of a pioneer in the field of nature conservation in Brazil. *Ambiente & Sociedade*, Campinas, No. 8, pp. 1-26, Jan/Jun 2005.

HENRIQUES, R.P.B. Influence of history, soil and fire on the distribution and dynamics of phytophysiognomies in the Cerrado biome. *In*: SCARIOT, A.; SOUZA-SILVA, J.C.;

FELFILI, J.M. (ed.). *Cerrado*: ecology, biodiversity and conservation. Ministry of the Environment, Brasília, pp. 73-92, 2005.

FERNANDEZ, F. Learning the lesson of Chaco Canyon: from "sustainable development" to sustainable living. *Reflexión*, n. 15, 2005. available at: <http://bit.ly/2smHfxc> Accessed on: 09 June 2017.

FERNANDEZ, F. *The imperfect poem*: chronicles of biology, nature conservation and its heroes. 3rd edition. Curitiba: UFPR, 2011.

FLORIANI, D. Sciences in transit, complex objects: socio-environmental practices and discourses. *Ambiente & Sociedade*, no. 9, pp. 65-80, 2006.

HENRIQUES, R.P.Barros. The threatened future of the Brazilian Cerrado. *Ciência*

Hoje, Rio de Janeiro, v. 33, n. 185, p. 34-39, jul. 2003.

HOEHNE, F.C. *As plantas ornamentaes da Flora Brasílica*. Vol. 1. São Paulo: Secretariat of Agriculture, Industry and Commerce, 1930.

KETTUNEN, M. *et al. Guidance on the maintenance of landscape connectivity features of major importance for wild flora and fauna*: guidance on the implementation of Article 3 of the Birds Directive (79/409/EEC) and Article 10 of the Habitats Directive (92/43/EEC). London, Institute for European Environmental Policy, 2007.

KLINK, C.A.; MACHADO, R.B. The conservation of the Brazilian Cerrado. *Megadiversidade*, Belo Horizonte, n. 1, pp. 147-155, Jul. 2005.

LAVRATTI, P.C.; PRESTES, V.B. *Diagnosis of legislation*: Identification of standards affecting mitigation and adaptation to climate change: deforestation/land use change, 2009. Available at <http ://bit.ly/2tffinE>. Retrieved 25 June 2010.

LEFF, E. *Epistemologia ambiental*. Translation by S. Valenzuela. 2nd edition. São Paulo: Cortez, 2002.

MACHADO, R.B. *et al.* Estimates of area loss in the Brazilian Cerrado. Brasilia: Conservation International (Brazil), 2004.

MACIEL, L.G. Efficiency and effectiveness of legal reserves and permanent protected areas in the Cerrado. Master's dissertation, Centre for Sustainable Development, University of Brasília, 2008.

MMA (Ministry of the Environment). *Action Plan for the Prevention and Control of Deforestation and Fires in the Cerrado*: Conservation and Development. Available at: <http://bit.ly/2tfbtPt>. Retrieved on: 21 May 2012.

MANTOVANI, W. The degradation of Brazilian biomes. *In*: RIBEIRO, W.C. (org.). Património ambiental brasileiro. (Coleção Uspiana - Brasil 500 anos). São Paulo: Edusp/Imprensa Oficial, 2003.

MARTINS JR, P.P. *Fundamental epistemology*: an introductory study of the structure of knowledge and the practical application of epistemology in scientific research. Science & Reality/Applied Epistemology Programme. Ouro Preto: UFOP/EM/DEGEO & CETEC, 2000. handout.

MARTIUS, C.F.P. von. A fisionomia do Reino Vegetal no Brasil [1824]. *Anuário Brasileiro de Economia Florestal*, Rio de Janeiro, n. 10, pp. 209-227, 1958.

METZGER, J.P. Is the Forest Code scientifically sound? *Natureza & Conservação*, São Paulo, n. 8, pp. 1-5, 2010.

MORAES, A.C.R. *Meio ambiente e Ciências Humanas*. 4th ed. ampl. São Paulo: Annablume, 2005.

MORIN, E. *Os sete saberes necessários à educação do futuro*. Translated by C.E.F. Silva; J. Sawaya. Technical revision by E.A. Carvalho. São Paulo: Cortez, 2001.

NABUCO, J. *O abolicionismo* [1883]. Available at: <http://bit.ly/2shsmvu>. Accessed on: 14 Apr. 2009.

NOVAES, W. *Cerrado*: a drama in silence. National Geographic (Brazil), São

Paulo, No. 190, pp. 54-67, Oct. 2008.

OST, F. *A natureza à margem da lei*: a Ecologia à prova do Direito. Lisbon: Piaget Institute, 1997.

PÁDUA, J.A. *The predatory heritage and its overcoming*. Available at: <http://brasilsustentavel.org.br/downloads.htm>. Accessed on: 13 Feb. 2008.

PÁDUA, J.A. The "monocultural spirit" and the authoritarian occupation of Brazilian territory. *Proposta*, Rio de Janeiro, n. 99, pp. 6-12, Dec/Feb, 2004a.

PÁDUA, J.A. *Um sopro de destruição*: pensamento político e crítica ambiental no Brasil escravista, 1786-1888. 2. ed. Rio de Janeiro: Zahar, 2004b.

PÁDUA, J.A. One country and six biomes: a conceptual tool for sustainable development and environmental education. *In*: PÁDUA, J.A. (org.) *Development, Equity and Environment*. Belo Horizonte/São Paulo: UFMG; Peirópolis, pp. 118-150, 2009.

QUINTELA, A.C. Do sertão ao Cerrado do Planalto Central: uma questão de nomenclatura. *Revista UFG*, Goiânia, n. 9, p. 242-257, dec. 2010.

RAMBALDI, D.M.; OLIVEIRA, D.A.S. (eds.). *Fragmentation of ecosystems*: Causes, impacts on biodiversity and recommendations for public policy. Brasília: Ministry of the Environment/Secretariat of Biodiversity and Forests, 2005.

SALOMON, M. The deforestation rate of the Cerrado is twice as high as that of the Amazon. *Folha de S. Paulo*, Cad. Brazil, 11 September 2009, p. A9.

SANTOS, B.S. Reinventing democracy: between pre-contractual law and post-contractual law. *Oficina do CES*, Coimbra, n. 107, Apr. 1998.

SANTOS, B.S. Uma cartografia simbólica das representações sociais: prolegómenos a uma concepção pós-moderna do Direito. *Revista Crítica de Ciências Sociais*, Coimbra, n. 24, pp. 139-172, Mar. 1988.

SANTOS, B.S. *Um discurso sobre as ciências*. 12th edition. Porto: Afrontamento, 2001.

SBPC (Brazilian Society for the Promotion of Science); ABC (Brazilian Academy of Sciences). *The Forest Code and Science*: Contributions to the Dialogue. 2nd ed, rev. São Paulo: SBPC, 2012. available at: <http://bit.ly/2rUMt1Z>. Retrieved on: 09 June 2017.

SAWYER, D. *PEC of the Cerrado and the Caatinga*: pros and cons. Available at: <http://bit.ly/2rcCYIE>. Retrieved on: 17 October 2008.

SILVA, C. R. (ed.). *Geodiversity in Brazil*: Knowing the past to understand the present and predict the future. Rio de Janeiro: CPRM, 2008.

STEIGLEDER, A.M. *Responsabilidade civil ambiental*: as dimensões do dano ambiental no direito brasileiro. Porto Alegre: Livraria do Advogado, 2004.

TABARELLI, M.; GASCON, C. Lessons from fragmentation research: improving management policies and guidelines for biodiversity conservation. *Megadiversidade*, Belo Horizonte, No. 1, pp. 181-188, Jul. 2005.

UNITED NATIONS (General Assembly). *Report of the World Commission on Environment and Development*: "Our Common Future", 1987. available at:

<http://bit.ly/2rcBwWK>. Retrieved on: 09 June 2017.

VAZ, H.C.L. *Mystical Experience and Philosophy in the Western Tradition*. São Paulo: Loyola, 2000.

VEIGA, J.E. *Sustainable development*: the challenge of the 21st century. 3rd edition. Rio de Janeiro: Garamond, 2008.

WALTER, B.M.T. Phytophysiognomies of the Cerrado biome: terminological synthesis and floristic relationships. 2006. 389 f. Dissertation (Doctorate in Ecology) - Institute of Biological Sciences, University of Brasília, Brasília, 2006.

WHITE JR, L.T. Historical roots of our ecological crisis. *Science*, Massachusetts, No. 155, pp. 1203-1207, 1967.

Table of contents

I **want** morebooks!

Buy your books fast and straightforward online - at one of world's fastest growing online book stores! Environmentally sound due to Print-on-Demand technologies.

Buy your books online at
www.morebooks.shop

Kaufen Sie Ihre Bücher schnell und unkompliziert online – auf einer der am schnellsten wachsenden Buchhandelsplattformen weltweit! Dank Print-On-Demand umwelt- und ressourcenschonend produzi ert.

Bücher schneller online kaufen
www.morebooks.shop

Printed by Books on Demand GmbH, Norderstedt / Germany